STUDENT GUIDE

FAMILY PORTRAITS

COMPARING FUNCTION FAMILIES

$y = -x^2$

MathScape
SEEING AND THINKING
MATHEMATICALLY

At the start of this phase, you will look at what it means for a mathematical relationship to be a function. Then you will collect and graph data on heel-to-toe pacing and data on the apparent size of an object. You will see how these sets of data relate to direct and inverse variation.

How can you describe mathematical relationships?

FAMILY
PORTRAITS

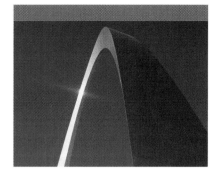

PHASE**TWO**
Linear Functions

In this phase you will look at functions whose graphs are straight lines. You will start by using a ratio to describe a line's slope. Next you will see how a line's equation can give you information about the line's slope and y-intercept. Finally, you will see how you can write the equation of a line that goes through two given points.

PHASE**THREE**
Quadratic Functions

How is the area of a slide's projected image related to the projector's distance from the screen? You will graph data to help you answer this question and learn how it connects to quadratic functions. You will also graph other quadratic functions and explore the relationship between their equations and their graphs.

PHASE**FOUR**
Exponential Functions

You will begin by thinking about the number of regions formed when a sheet of paper is folded in half over and over. You will graph this exponential function and explore some rules of exponent arithmetic. Finally, you will work with a calculator to see how scientific notation can help you write very large and very small numbers.

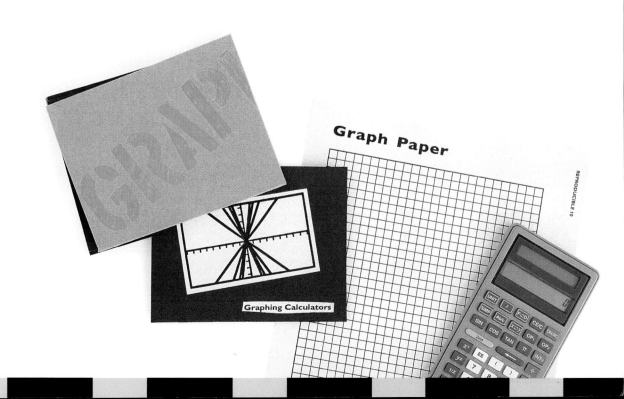

Graph Paper

Graphing Calculators

PHASE ONE

In this phase you will see what it means for a mathematical relationship to be a function. You will also do some experiments to help you investigate two types of functions—direct and inverse variation.

You will be collecting data on the relationship between the number of heel-to-toe paces you take and the distance you cover. What other mathematical relationships have you worked with in the past?

Direct and Inverse Variation

WHAT'S THE MATH?

Investigations in this section focus on:

ALGEBRA

- Understanding what makes a mathematical relationship a function
- Recognizing functions from their graphs
- Graphing and describing direct variation functions
- Graphing and describing inverse variation functions

GEOMETRY and MEASUREMENT

- Collecting measurement data
- Displaying and analyzing measurement data

MathScape Online
mathscape3.com/self_check_quiz

1 Inputs and Outputs

Take any number, double it, and add 1. This rule tells you how to take any input value and get an output value. Can you think of other rules that relate inputs and outputs? You will look at examples of such rules, and explore a special type of rule called a function.

Find Each Machine's Rule

How can you write rules to describe the relationship between an input and an output?

In each picture, a machine takes input values and gives you output values. Each machine has a printout showing some sample inputs and outputs.

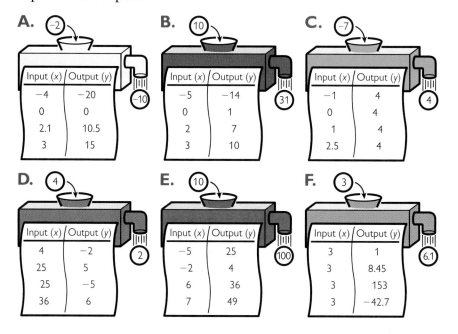

A. (−2)

Input (x)	Output (y)
−4	−20
0	0
2.1	10.5
3	15

(−10)

B. (10)

Input (x)	Output (y)
−5	−14
0	1
2	7
3	10

(31)

C. (−7)

Input (x)	Output (y)
−1	4
0	4
1	4
2.5	4

(4)

D. (4)

Input (x)	Output (y)
4	−2
25	5
25	−5
36	6

(2)

E. (10)

Input (x)	Output (y)
−5	25
−2	4
6	36
7	49

(100)

F. (3)

Input (x)	Output (y)
3	1
3	8.45
3	153
3	−42.7

(6.1)

1. For each machine, describe the relationship between the input and the output. Whenever possible, write an equation that describes the machine's rule.

2. Write a brief description of how Machines A, B, D, and E are different from Machines C and F.

Decide If It's a Function

A mathematical rule that assigns exactly one output value to each input value is a **function.** Each of the following graphs shows a relationship. The input values are along the horizontal axis; the output values are along the vertical axis.

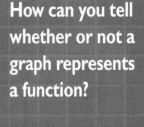
How can you tell whether or not a graph represents a function?

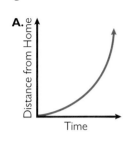

A. Distance from Home / Time

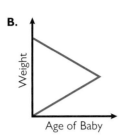

B. Weight / Age of Baby

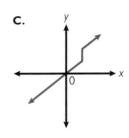

C.

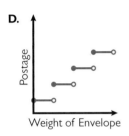

D. Postage / Weight of Envelope

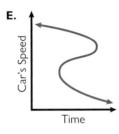

E. Car's Speed / Time

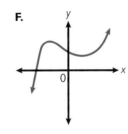

F.

1 Which graphs show functions? Why?

2 How were you able to tell which graphs show functions? State a general rule you can use to tell whether or not a graph is a function.

Set Up a Function Album

In this unit you will be investigating different types of functions and recording facts in a function album. For the first page of your album, write a summary of what you know so far about functions. Include the following:

- the definition of a function, in your own words

- examples of functions

- how you can tell whether a graph shows a function

hot words | function
rule

Homework

page 302

2 Keeping a Steady Pace

How long is your classroom? You could measure it directly with a tape measure, but it is often easier to measure long distances indirectly. You will be finding a way to measure this length by pacing. You will also see how finding distances by pacing is related to a type of function known as direct variation.

Make a Conversion Graph

How can you make a graph to show the relationship between paces and distance?

Your teacher has set up some marks on the ground that are one meter apart.

1 Starting at the first mark, begin pacing "heel-to-toe." Record various numbers of paces and their distances in meters.

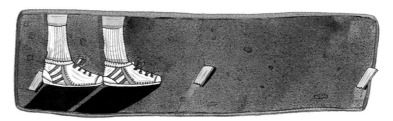

2 Plot your data to make a conversion graph that relates the number of paces to the distance covered.

3 Write an equation for the graph. Let *x* be the number of paces and *y* be the distance in meters. Your equation should help you convert a given number of paces into distance covered.

Find the Length of Your Classroom

Carefully pace off the length of your classroom. Then use your graph and/or equation to find the length of the classroom. Write a summary of your findings that includes the following:

- The length of your classroom
- A complete description of your method

Explore the $y = kx$ Family

Equations that have the form $y = kx$ (where k is not equal to 0) are called **direct variation** functions. What do the graphs of these equations have in common?

What can you say about the graphs of equations that have the form $y = kx$?

1 Work with a partner to write down at least 8 equations of the form $y = kx$. Use both positive and negative values for k. Also, choose some values of k that are between 0 and 1.

2 Graph each of your equations on the same coordinate plane.

3 What do you notice about the graphs? Write a list of as many generalizations as possible. Here are some things to consider.

  a. What do all the graphs have in common?

  b. Where do the graphs intersect?

  c. Which graphs slant upward as you move from left to right? downward?

  d. Which graphs are steepest? flattest?

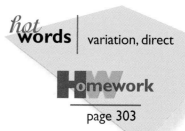

hot **words** | variation, direct

Homework

page 303

3 How Long Is a Meter?

Have you ever noticed how objects appear to get smaller as you move farther away from them? You will collect data involving the "apparent size" of a meter. Then you will see how this is related to inverse variation.

Graph Apparent Size Versus Distance

How is the apparent size of an object related to your distance from it?

Work with a partner for this investigation. Decide which of you will be Partner A and which will be Partner B.

1 Partner A: stand with your back to the meter-long strip of paper that your teacher has hung. Then take at least 8 heel-to-toe paces in a straight line away from the strip.

2 Partner A: turn around and use a ruler to measure the apparent size of the strip to the nearest tenth of a centimeter.

3 Partner B: record the number of paces, *x*, and the apparent size, *y*, in the table on the Apparent Size Recording Sheet.

4 Repeat steps 1–3 with different numbers of paces. Collect at least 10 data points.

Plot your points using the recording sheet's axes. What is your graph's shape? Find $x \cdot y$ for each pair of values. What do you notice?

Measuring Apparent Size

To measure the apparent size of an object:

- Stand facing the object.
- Hold your arm straight out.
- Hold a ruler vertically aligned with the object.
- Read the apparent length of the object.

Explore the $y = \frac{k}{x}$ Family

Equations that have the form $y = \frac{k}{x}$ or $xy = k$ (where k is not equal to 0) are called **inverse variation** functions. What do the graphs of these equations have in common?

Your teacher will assign your group a value of k to work with.

1 Make a table of values for your equation.

    a. Include at least 10 pairs of points with positive values of x.

    b. Include at least 10 pairs of points with negative values of x.

    c. Include some values of x that are very close to 0.

2 Plot these points to help you graph the equation.

Compare your graph to those of other groups.

What can you say about the graphs of equations that have the form $y = \frac{k}{x}$?

Update the Function Album

Add two pages to your function album—one that summarizes what you know about direct variation functions, and one that summarizes what you know about inverse variation functions. Include the following on each page:

- a verbal description of the function family

- a typical graph from the function family

- a sample equation for the function family

- an example of how the function occurs in a real-world situation

- anything else that might be a useful reference about the function family

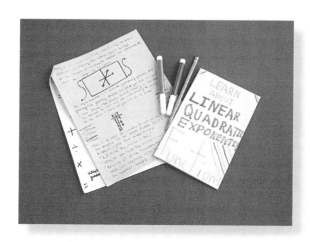

hot **words** | variation, inverse

Homework

page 304

PHASE TWO

Linear functions have graphs that are straight lines. In this phase, you will take a close look at the ideas of slope and y-intercept. You will see how these concepts are connected to the equation of a linear function.

A train moving at a constant speed has a distance-versus-time graph that is a straight line. A graph showing the number of paces you take versus the distance you cover is also a straight line. What other relationships have you seen that have linear graphs?

Linear Functions

WHAT'S THE MATH?

Investigations in this section focus on:

ALGEBRA

- Finding the slope of a line
- Finding the y-intercept of a line
- Connecting the slope and y-intercept of a line to its equation
- Writing the equations of lines based on their graphs

RATIO and PROPORTION

- Using ratios to describe slopes

MathScape Online
mathscape3.com/self_check_quiz

4 A New Slant on Linear Functions

A linear function has a graph that is a straight line. Now it's time to think about how you can describe the slant, or slope, of a line. You will explore this idea and then see what it means for a line to have a negative slope.

Explore the Slope of a Line

How can you use math to describe the steepness, or slope, of a line?

Work with a partner for the following.

1 Start at any point on Line A and move horizontally as directed.

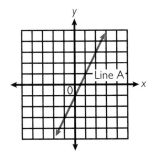 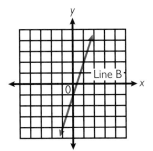

a. Move 1 unit to the right. How many units does the line rise? What result do you get if you start at any other point on the line?

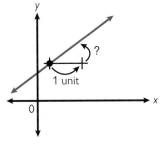

b. Move 2 units to the right. How many units does the line rise? What result do you get if you start at any other point on the line?

c. Move 3 units to the right. How many units does the line rise? What result do you get if you start at any other point on the line?

2 Repeat the same moves with Line B. What results do you get?

3 Write a brief statement about what you have noticed. Be ready to share your results with the class.

Find Slopes of Some Lines

For each pair of points, plot the points and draw the line through them. Then calculate the slope of the line.

1 $(-1, 4)$ and $(2, 1)$

2 $(1, 6)$ and $(6, 4)$

3 $(-4, 2)$ and $(2, -3)$

4 $(3, 2)$ and $(7, 2)$

5 $(-1, -1)$ and $(5, -1)$

What do all of the lines in 1–3 have in common? What do you notice about their slopes? What do the lines in 4 and 5 have in common? What do you notice about their slopes?

What can you say about the slope of lines that slant downward from left to right?

How to Find the Slope of a Line

Choose two points along the line. The **rise** is the difference in the y-coordinates. The **run** is the difference in the x-coordinates.

The **slope** of the line is the ratio of the rise to the run:

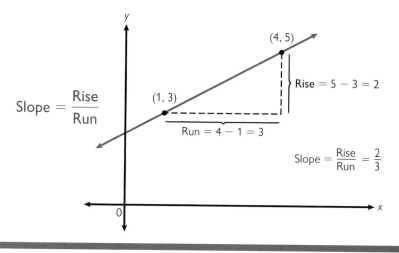

$$\text{Slope} = \frac{\text{Rise}}{\text{Run}}$$

(4, 5)

Rise $= 5 - 3 = 2$

(1, 3)

Run $= 4 - 1 = 3$

$$\text{Slope} = \frac{\text{Rise}}{\text{Run}} = \frac{2}{3}$$

hot words | rise
run
slope

Homework
page 305

5 What's in an Equation?

Did you know that the equation of a line is like a coded message? You will be experimenting with the graphs of some lines. This will help you see how information about the slope and y-intercept of a line is contained in its equation.

Connect Equations, Slopes, and Intercepts

How are the slope and y-intercept of a line related to its equation?

Your teacher will give you and a partner two equations to work with. For each equation, do the following.

1 Make a table of values that satisfy the equation.

2 Plot the points from your table to help you graph the equation.

3 Find the slope of your graph.

4 Find the y-intercept of your graph.

For each equation, how are the slope and y-intercept of the graph related to the equation? What connections do you see?

The y-Intercept of a Line

The **y-intercept** of a line is the point where it crosses the y-axis.

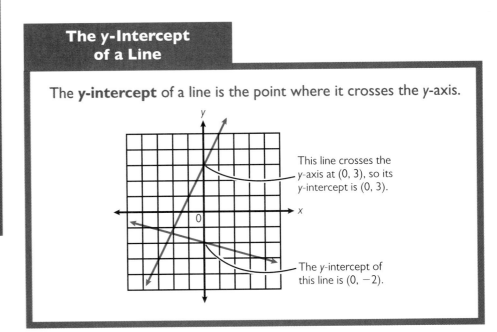

This line crosses the y-axis at (0, 3), so its y-intercept is (0, 3).

The y-intercept of this line is (0, −2).

Analyze a Set of Equations

Consider the set of six equations shown below.

A. $y = 5x$ B. $y = \frac{1}{3}x - 2$ C. $y = -x$

D. $y = 3x + 4$ E. $y = 0.5x + 1$ F. $y = -2x - 4\frac{1}{2}$

Work with classmates to write a response for each of the following. Be prepared to discuss your responses with the class.

1 Which equation has the flattest graph? How do you know?

2 Which equations have graphs that slant downward as you go from left to right? How do you know?

3 Which equation has a graph that crosses the y-axis at the highest point? Why?

4 Write an equation of your own whose graph would be steeper than the graph of any of the six given equations.

5 Write an equation of your own whose graph would be flatter than those of the six given equations.

6 Write an equation of your own whose graph would cross the y-axis at a lower point than would any graph of the six given equations.

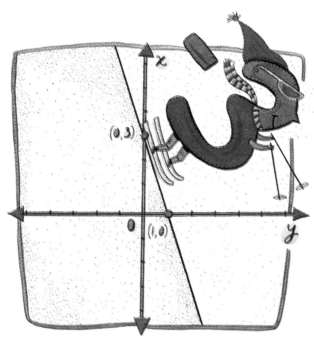

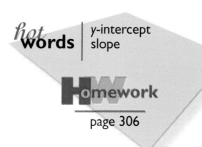

hot words | y-intercept | slope

H omework
page 306

6 The Shortest Distance Between Two Points...

WRITING THE
EQUATION OF
A LINE

Until now, you have usually started with the equation of a line and then created its graph. Now suppose you have the graph of a line and want to find its equation. You will be starting with two points, drawing the line through the points, and finding a way to write the line's equation.

Write the Equation of a Line

How can you write the equation of a line that goes through two points?

For this investigation, you will need a copy of the reproducible Starting Points.

1 Choose any two of the five given points on the coordinate plane. Using a straightedge, carefully draw a line through the points.

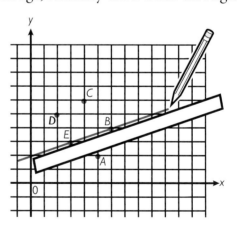

2 Is the slope of your line positive, negative, or zero? Is the y-intercept positive, negative, or zero? Based on your answers, what can you say about the equation of your line? What do you think your equation will look like?

3 Calculate the slope of your line.

4 Find the y-intercept of your line.

5 Use the information you have gathered to write the equation of your line.

Check the Equation

Refer to your line and the equation you wrote for it. Write a brief response to each of the following items.

How can you be sure the equation you wrote is correct?

1 According to your graph, give the coordinates of at least three points that lie on your line.

2 Since these points lie on the line, how do their coordinates relate to the line's equation?

3 Use this fact to see if your equation is correct. If your equation is not correct, look for errors in your work and revise your equation.

Update the Function Album

Add a page to your function album that summarizes what you know about linear functions. Include the following:

- a verbal description of this family of functions

- a typical graph from the function family

- a sample equation for the function family

- a description of how to calculate slope

- a description of how to find the equation of a line through two points

- anything else that might be a useful reference about the function family

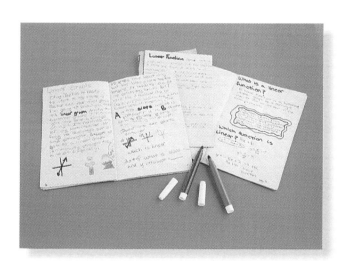

hot**words** | y-intercept
slope

Homework

page 307

PHASE THREE

In this phase, you will look at functions whose graphs are parabolas. These are called quadratic functions.

The first situation you will explore involves the area of a slide projector's image. By collecting data, graphing it, and writing an equation, you will see how this situation relates to quadratic functions. You will also see how a real-world problem about rectangular pens for animals results in a quadratic function.

Quadratic Functions

WHAT'S THE MATH?

Investigations in this section focus on:

ALGEBRA

- Writing equations for quadratic functions
- Graphing parabolas
- Understanding how the constant a in the equation $y = ax^2$ affects the graph of the equation
- Using quadratic functions to solve problems

PATTERNS

- Describing and extending patterns that result from quadratic functions

MathScape Online
mathscape3.com/self_check_quiz

7 The Area of a Projected Image

EXPLORING A
QUADRATIC
RELATIONSHIP

How is the area of a slide's projected image related to the projector's distance from the screen? You will see how this relationship—called a quadratic function—can be described with a graph and an equation.

Make a Graph of Area Versus Distance

What does the graph of this area-versus-distance relationship look like?

The figure below shows the relationship between the area of a projected image and the projector's distance from the screen.

1 Make a table of values that relates the area of the image to the distance of the projector from the screen.

Distance from screen (m)	Area of image (cm²)
0	0
1	
2	
3	
4	
5	

2 Plot the points from your table to help make a graph of the relationship.

How is the graph of this relationship different from the graph of a linear function?

Area Versus Distance

When a square slide is projected onto a screen, the area of the image depends upon the distance of the projector from the screen.

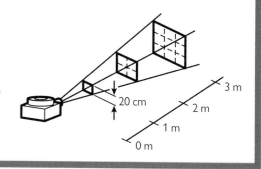

Write an Equation for the Relationship

Use your table to help you write an equation that describes the relationship between the area of a projected image and the distance of the projector from the screen. Given any distance, d, your equation should tell how to calculate the area, A.

What equation can you write to describe the area-versus-distance relationship?

Write About the Results

Write a brief summary of your findings. Include the following:

- a description of how your equation is different from the equation of a linear function

- a description of how your graph is different from the graph of a linear function

- a discussion of how the area of the image changes as the distance changes

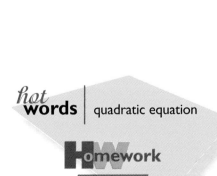

hot **words** | quadratic equation

H**mework**

page 308

Going Around a Curve

What does the graph of a quadratic function look like?

You have already seen one quadratic relationship: the area of a slide's projected image as a function of the projector's distance from the screen. Now you will explore the graphs of other equations that contain an x^2 term.

Make and Describe a Graph

What does the graph of the function $y = x^2$ look like?

Equations of the form $y = ax^2 + bx + c$, where $a \neq 0$, are called **quadratic functions.** The simplest of these is $y = x^2$.

1 Make a table of values for the equation $y = x^2$. Be sure to include at least four negative values of x.

2 Use your table to plot points and make a graph.

The graph you made is called a **parabola.** Write a description of your graph. Include the ideas of symmetry and axis of symmetry in your description.

Symmetry

If you can fold a figure in half so that the two halves match perfectly, then the figure has **symmetry**. The fold line is called the **line of symmetry** or the **axis of symmetry**.

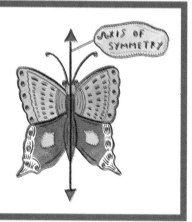

Experiment with Parabolas

Work with classmates to make a graph for each of the six equations shown. Then be ready to discuss the questions below with the rest of the class.

How is the shape of a parabola related to its equation?

Column A	Column B
$y = x^2$	$y = -x^2$
$y = 2x^2$	$y = -2x^2$
$y = \frac{1}{2}x^2$	$y = -\frac{1}{2}x^2$

1 How do the graphs of the equations in Column A compare to those in Column B? What is similar? What is different? Include a discussion of symmetry in your answer.

2 What does the constant a in the equation $y = ax^2$ tell you about the graph?

3 Answer this question without actually making any graphs: How does the graph of $y = 3x^2$ compare to the graph of $y = 5x^2$?

hot words | parabola symmetry

Homework

page 309

 **Fenced In**

Suppose you have 32 m of fencing material. What is the largest rectangle you can fence off? And what does this problem have to do with parabolas? You will explore this situation and make a graph to describe it. This will help you see the connection between perimeters, areas, and parabolas.

How can you find a rectangle with the greatest area for a given perimeter?

Explore Rectangular Pens

A farmer has 32 m of fencing material and wants to fence off a rectangular pen for animals. One side of the pen must lie along a creek. What length along the creek results in a pen with the greatest area for the animals?

1 What is the perimeter of any pen the farmer can make?

2 Use a sheet of graph paper to help sketch all of the possible pens that have whole-number lengths. One possible pen is shown here.

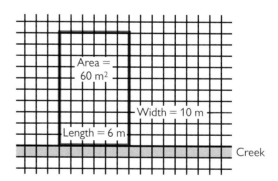

3 Find the area of each pen.

Which length along the creek results in the pen with the greatest area?

Plot the Pens

Make a graph that shows the relationship between the length of the fence along the creek and the area of the rectangular pen.

How can a graph of the situation help solve the problem?

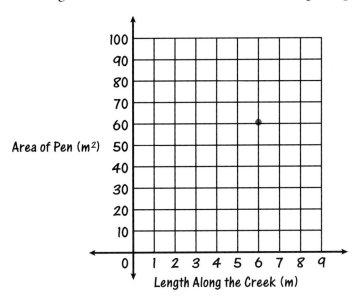

Area of Pen (m²)

Length Along the Creek (m)

1 Describe your graph.

2 What is the line of symmetry of your graph?

3 From your graph, what length along the creek results in a pen with the greatest area? Does this agree with your earlier results?

Update the Function Album

Add a page to your function album that summarizes what you know about quadratic functions. Include the following:

- a verbal description of the function family

- sample equations for the function family

- a typical graph from the function family and a discussion of symmetry

- a discussion of how the constant a affects the graph of $y = ax^2$

- anything else that might be a useful reference about the function family

hot **words** | area \ function

H🔲🔲mework

page 310

PHASE FOUR

The more times you fold a sheet of paper in half, the more rectangular regions there are when unfolded. How is the number of regions related to the number of folds?

You may be surprised when you collect data to help answer this question. Exponential functions have some special characteristics. In this phase, you will explore laws of exponents. You will also see how your calculator handles very large and very small numbers.

Exponential Functions

WHAT'S THE MATH?

Investigations in this section focus on:

ALGEBRA

- Graphing exponential functions
- Writing equations for exponential functions

PATTERNS

- Describing and extending patterns that result from exponential functions

NUMBER

- Developing and using laws of exponents
- Working with scientific notation

MathScape Online
mathscape3.com/self_check_quiz

10 Folds and Regions

EXPLORING
EXPONENTIAL
FUNCTIONS

Imagine folding a sheet of paper in half 25 times. If you unfolded the sheet of paper, how many rectangular regions would be formed? You will explore this question and see how it is connected to exponential functions. Then you will compare the graphs of several exponential functions.

Calculate the Number of Regions

What is the relationship between the number of times a sheet of paper is folded and the number of regions formed?

When you fold a sheet of paper in half one time, two regions are formed. If you fold the sheet of paper in half again, four regions are formed.

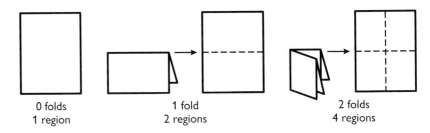

0 folds
1 region

1 fold
2 regions

2 folds
4 regions

1 Make a table that relates the number of folds to the number of regions. (Save your table for use later in this unit!)

2 Describe the relationship between the number of folds and the number of regions using words, variables, or any other method that makes sense to you.

3 Suppose a sheet of paper could be folded in half 25 times. Find the number of regions that would be formed.

Compare Graphs of Exponential Functions

Functions of the form $y = a^x$ are called **exponential functions.** Work with classmates to compare the exponential functions $y = 3^x$, $y = 4^x$, and $y = 5^x$.

1 Make a table of values for each function. Include $x = 0, 1, 2, 3, 4, 5,$ and 6. One way to organize your tables is shown here. (Save your tables for use later in this unit!)

x	3^x	4^x	5^x
0	$3^0 = 1$	$4^0 = ?$	
1	$3^1 = 3$		
2			
3			
4			
5			

2 Make graphs for all three functions on the same set of axes.

Write a brief description of what all three graphs have in common. What are some differences?

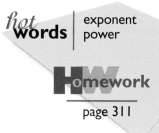

hot **words** | exponent power

Homework
page 311

11 Rules of the Road for Exponents

DEVELOPING LAWS OF EXPONENTS

How can you evaluate an expression that contains more than one exponent? As you will see, the tables of values you have already developed can be quite useful in exploring laws of exponents. You will also see how one of these laws can help you make more complete graphs of exponential functions.

Evaluate Expressions with Exponents

How can tables of values help you evaluate expressions with exponents?

Here is an example of how a table can help you evaluate $2^3 \cdot 2^4$ and translate the result back to an exponent.

$$2^3 \cdot 2^4 = 8 \cdot 16 = 128 = 2^7$$

2^2	4
2^3	8
2^4	16
2^5	32
2^6	64
2^7	128

Use the tables you have already made to help you evaluate each expression. Then use your tables to translate your result back into an exponent. Look for patterns as you work.

1. $3^2 \cdot 3^3$

2. $2^7 \cdot 2^{12}$

3. $2^{10} \cdot 2^6$

4. $\dfrac{2^{15}}{2^9}$

5. $\dfrac{4^5}{4^3}$

6. $\dfrac{2^{21}}{2^3}$

7. $(2^5)^4$

8. $(2^6)^3$

9. $(5^2)^2$

Write About Laws of Exponents

Write a summary of any patterns you noticed in the previous investigation.

- In general, what is true about $a^n \cdot a^m$?

- In general, what is true about $\dfrac{a^n}{a^m}$?

- In general, what is true about $(a^n)^m$?

298 FAMILY PORTRAITS • LESSON 11

Graph an Exponential Function

How can negative exponents help you graph exponential functions?

Negative Exponents

If a is a positive number, then $a^{-n} = \dfrac{1}{a^n}$.

For example, $4^{-3} = \dfrac{1}{4^3} = \dfrac{1}{4 \cdot 4 \cdot 4} = \dfrac{1}{64}$.

Now that you know how to work with negative exponents, you can graph exponential functions for both positive and negative values of x. Your teacher will give you a function to work with.

1 Make a table of values for your function. Be sure to include at least three negative values for x, $x = 0$, and at least three positive values for x.

2 Plot the points to help make a graph of your equation.

Write a brief description of the shape of your graph.

hot **words** | exponent power

Homework
page 312

12 The Very Large and the Very Small

You have seen how quickly the function $y = 2^x$ grows as x increases. What is the largest value of 2^x that your calculator can handle? What is the smallest? As you explore scientific notation, you will see how it can help you answer these questions about your calculator.

Make a Powers-of-10 Table

What patterns do you notice in a table of powers of 10?

You will need a copy of the Powers-of-10 Table. Fill in as much of the table as you can. Work with students around you to help with parts of the table you are unsure about.

Power of Ten	Number	Common Name of Number	Power of Ten	Number (written as a fraction)	Number (written as a decimal)	Common Name of Number
10^0						
10^1			10^{-1}			
10^2			10^{-2}	$\frac{1}{100}$	0.01	one one-hundredth
10^3			10^{-3}			

Be ready to discuss the following questions with the class.

- What patterns do you notice in your table?

- How can you tell how many zeros are in a number based on the power of 10?

Scientific Notation

Numbers written in **scientific notation** have the form $a \times 10^n$, where a is a number between 1 and 10, and n is an integer. Here are some examples:

$4,000 = 4 \times 10^3$

$320,000,000 = 3.2 \times 10^8$

$0.0007 = 7 \times 10^{-4}$

Explore a Calculator's Limits

Find the greatest power of 2 you can enter into your calculator without causing it to display an error message.

1 Experiment by entering various powers of 2 into your calculator using the y^x or $\land$ key.

2 How does your calculator display this largest power of 2?

3 How would you write this number using scientific notation?

4 If you wrote out this number in full, how many digits would it have?

Repeat the above process to find the smallest (negative) power of 2 you can enter into your calculator.

How does your calculator display very large and very small numbers?

Update the Function Album

Add a page to your function album that summarizes what you know about exponential functions. Include the following:

- sample equations for the function family
- a typical graph from the function family
- a summary of laws of exponents
- a description of how to write numbers in scientific notation
- anything else that might be a useful reference about the function family

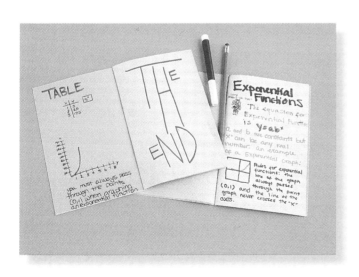

hot **words** | power
scientific notation

Home**work**

page 313

Inputs and Outputs

Applying Skills

A machine takes input values and gives output values. Some sample inputs and outputs are shown. Describe each machine's rule using words and/or an equation.

1.

Input (x)	Output (y)
−2	−6
0	0
1	3
2.2	6.6

2.

Input (x)	Output (y)
−3	9
0	0
1	1
2	4

3.

Input (x)	Output (y)
−3	−2
0	−2
2	−2
6	−2

4.

Input (x)	Output (y)
9	3, −3
16	4, −4
81	9, −9
100	10, −10

5. Which rules in items 1–4 are functions? Why?

Tell whether each graph shows a function and explain your thinking.

6.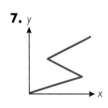

7.

8.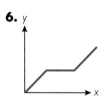

9.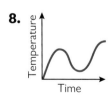

Extending Concepts

10. Is the relationship $y = \pm\sqrt{x}$ a function? Is the relationship $y = x^2$ a function? Explain your thinking.

11. Describe two real-world examples of functions. How do you know these relationships are functions?

12. Sketch a graph of your own that represents a function and a graph that does not represent a function. How can you tell whether a graph represents a function?

Making Connections

13. Suppose that a weight is attached to the end of a swinging pendulum. As the pendulum swings, the distance of the weight from the center of the pendulum's arc varies as shown in the graph.

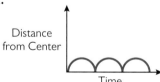

Is the relationship between distance and time a function . . .

a. if time is the input value and distance from the center is the output value?

b. if distance from the center is the input value and time is the output value?

Explain your answers.

Keeping a Steady Pace

Applying Skills

Graph each equation on the same coordinate plane.

1. $y = 2x$

2. $y = 0.4x$

3. $y = -3x$

4. $y = 4x$

Tell whether each graph represents an equation of the form $y = kx$. If it does, tell whether k is positive or negative. If not, say why not.

5.

6.

7.

8.
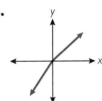

9. Where do the graphs of $y = 8x$ and $y = -4x$ intersect?

10. Which equation below has the steepest graph? the flattest graph? How do you know?

$y = -x$ $y = 5x$

$y = 0.5x$ $y = -6x$

11. Which equations below have graphs that slant downward from left to right? How do you know?

$y = x$ $y = -8x$

$y = -0.1x$ $y = 2x$

Extending Concepts

Tell whether each function is a direct variation function. If it is, write an equation describing the relationship. If it is not, explain why not.

12. A person's salary as a function of the number of hours she works (assume she makes $15 per hour)

13. The number of legs as a function of the number of dogs

14. The cost of a rental car as a function of the number of miles driven (the car costs $20 plus 10 cents per mile)

15. The height of a child as a function of his or her age

16. The revenue for a show as a function of the number of tickets sold (tickets cost $18 each)

Making Connections

17. At maximum speed, a cheetah can run about 100 feet per second. The giant tortoise can cover about 0.25 feet per second. For each animal, write an equation relating distance and time. Assume each animal is moving at its maximum speed. What do the graphs of these equations look like? How are the graphs alike? How are they different?

How Long Is a Meter?

Homework

Applying Skills

1. a. Make a table of values for the function $y = \frac{12}{x}$. Include $x = -24, -12, -6, -4, -3, -1, -0.5, 0.5, 1, 3, 4, 6, 12, 24$.

   b. Plot your points and make a graph of the equation.

Tell whether each of the following could be the first-quadrant graph of the equation $y = \frac{k}{x}$ (k is a positive number). Explain why or why not.

2.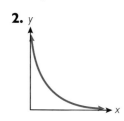

3.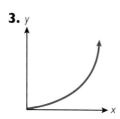

4.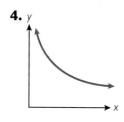

5.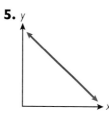

Tell whether each equation represents direct or inverse variation.

6. $y = 80x$ **7.** $xy = 80$

8. $y = \dfrac{80}{x}$ **9.** $\dfrac{y}{x} = 80$

Tell whether each statement about the graph of $y = \frac{k}{x}$ is true or false. If it is false, change it to make it true. Assume that k is positive.

10. As x increases in the first quadrant, the graph rises steeply.

11. The graph appears only in the first and second quadrants.

12. In the first quadrant, the graph rises steeply as x gets closer to zero.

13. As x gets very large, the graph will eventually cross the x-axis.

Extending Concepts

14. Jim makes $10 per hour. His total pay is a function of the number of hours he works. Is this an inverse variation function? Why or why not?

15. Suppose you drive at constant speed. The time (in hours) it takes to drive 500 miles and your speed (in mph) are related by the equation $t = \frac{500}{s}$.

   a. Make a graph of this equation. Show s on the horizontal axis.

   b. What happens to your graph as s gets very large? What happens as s gets very close to zero? Explain why your observations make sense.

Writing

16. Write a short summary about the graphs of direct variation and inverse variation functions. Describe how the graphs of the two types of functions differ.

A New Slant on Linear Functions

Applying Skills

Find the slope of each line.

1.
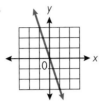

2.

For each pair of points, plot the points and draw the line through them. Then calculate the slope of the line.

3. $(1, 2)$ and $(3, 6)$

4. $(2, 6)$ and $(4, 0)$

5. $(-3, 2)$ and $(1, 4)$

6. $(2, 5)$ and $(1, 5)$

7. $(-2, -2)$ and $(0, -5)$

8. $(-4, 1)$ and $(-6, -3)$

9. If a line is horizontal, what is its slope? Why?

10. If the slope of a line is negative, does it slope upward or downward as you move from left to right? Why?

11. If the slope of a line is 4, how many units does the line rise if you move 2 units to the right?

12. If the slope of a line is 5, what rise corresponds to a run of -3?

13. If the slope of a line is -6, what run corresponds to a rise of 12?

Extending Concepts

14. A line has slope $-\frac{1}{2}$, and the point $(1, 4)$ lies on the line. Find the coordinates of a second point that lies on the line and graph the line.

15. Explain why the slope of a vertical line is undefined.

Making Connections

16. *Apollo 10*, which orbited the moon, was launched in 1969. The graph represents one portion of its journey and shows the distance traveled as a function of time.

a. Choose two points on the graph and write their coordinates. Find the rise and run corresponding to this pair of points. What does the rise represent? What does the run represent?

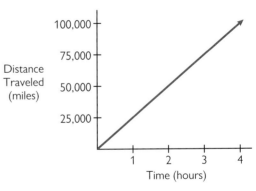

b. What is the slope of the graph? What does it represent?

What's in an Equation?

Applying Skills

Find the y-intercept of each line.

1.   **2.**

Find the slope and y-intercept of the graph of each equation.

3. $y = 5x + 8$

4. $y = -2x + 1$

5. $y = x - 4$

6. $y = 0.3x - 2.5$

7. $y = -3x$

8. $y = \frac{1}{4}x + \frac{2}{5}$

9. $y = 2 - x$

10. $y = 2$

11. Which equation in items **3–10** has the steepest graph? the flattest graph?

12. Which equations in items **3–10** have graphs that slant upwards from left to right?

Write the equation of a line with the following slope and y-intercept:

13. slope 7, y-intercept $(0, -9)$

14. slope -4, y-intercept $(0, 1)$

15. slope 4.5, y-intercept $(0, 0)$

16. slope -6, y-intercept $(0, -1)$

Extending Concepts

Find the slope and y-intercept of the graph of each equation.

17. $2y = 4x + 6$

18. $x + y = 7$

19. $y = 5(x + 1)$

20. $y - 2x = 4$

In items **21–23**, use the equations below.

A. $y = 4x + 2$ **B.** $y = \frac{1}{2}x - 3$

C. $y = -2x + 4$

21. Use the slope and y-intercept to graph each equation. Label the lines A, B, and C.

22. Write an equation whose graph crosses the y-axis at a lower point than all of the given lines.

23. Write an equation whose graph crosses the y-axis at a higher point than Line A and that is steeper than all three graphs.

Making Connections

24. Fahrenheit and Celsius are different temperature scales. The equation $F = 1.8C + 32$ describes the relationship between Fahrenheit and Celsius.

a. What is the slope of the graph of this equation? How much does the Fahrenheit temperature increase when the Celsius temperature increases by 1°? by 2°? How do you know?

b. What is the y-intercept of the graph? What is the Fahrenheit temperature when the Celsius temperature is 0°?

The Shortest Distance Between Two Points . . .

Applying Skills

1. a. Plot the points $(1, 4)$ and $(3, 0)$ on a coordinate plane and draw the line through the points.

b. Is the slope of the line positive, negative, or zero?

c. Is the y-intercept of the line positive, negative, or zero?

d. What can you say about the equation of the line?

e. Calculate the slope of the line.

f. Use your graph to find the y-intercept of the line.

g. Write the equation of the line.

h. Use your graph to find the coordinates of three new points that lie on the line. Check that the coordinates of each point satisfy the equation that you wrote. Show your work.

2. Repeat item **1** using the points $(1, 2)$ and $(2, 5)$.

3. Repeat item **1** using the points $(2, -1)$ and $(6, 1)$.

Extending Concepts

4. Find the equation of the line shown here.

Writing

5. Answer the Dr. Math letter.

> Dear Dr. Math,
>
> I wanted to find the equation of the line through the points $(-1, 0)$ and $(3, 1)$. I plotted the points and drew the line as I've shown. I calculated the slope this way: rise $= 1 - 0 = 1$, run $= 3 - (-1) = 4$, slope $= \frac{rise}{run} = \frac{1}{4}$. The y-intercept looked like about $\frac{1}{3}$, so I wrote this equation for the line: $y = \frac{1}{4}x + \frac{1}{3}$. But when I checked, I found that the coordinates $(3, 1)$ didn't satisfy this equation. I guess the y-intercept wasn't quite right after all. How can I figure out exactly what the y-intercept is?
>
> Mario

The Area of a Projected Image

Applying Skills

Tell whether or not each function is a quadratic function and why.

1. $y = 3x + 10$

2. $y = 5x^2$

3. $y = \dfrac{3}{x}$

4. $y = 8 - x$

5. $y = 0.1x^2$

For each table of values, write an equation that describes the relationship between the two variables. Then complete the table.

6.

x	y
0	0
1	3
2	12
3	27
4	?

7.

s	t
0	0
1	8
2	32
3	72
4	?

8.

p	q
0	0
1	−2
2	−8
3	−18
4	?

Extending Concepts

9. a. Make a table of values that relates the side length of a square (measured in yards) and its area (measured in square feet). A table has been started for you.

1 yd

1 ft
1 ft
1 ft

2 yd

Side Length (yd)	Area (ft)^2
0	0
1	9
2	
3	
4	
5	

b. Write an equation that describes the relationship between the side length, s, and the area, A.

c. How is your equation different from a linear equation?

d. Predict what the graph of your equation will look like. How will it differ from the graph of a linear equation?

e. Plot the points from your table and make a graph of the relationship.

Making Connections

10. Suppose that a rock is dropped from a tall building. Let d represent the distance (in feet) from the point of release and t the time (in seconds) since the rock was released. The variables d and t are related by the equation $d = 16t^2$.

a. Is this equation linear or quadratic? How do you know?

b. Make a table of values that satisfy the equation. Use values 0 through 5 for t.

c. Plot your points and make a graph of the equation.

d. How does the distance that the rock falls during the first second compare with the distance that it falls during the fifth second? Explain your thinking.

Going Around a Curve

Applying Skills

Make a table of values for each equation. Include negative as well as positive values of *x*. Use your table to plot points and make a graph.

1. $y = x^2$

2. $y = 3x^2$

3. $y = 0.5x^2$

4. $y = -2x^2$

5. $y = -0.25x^2$

6. $y = -4x^2$

The graphs shown here represent the equations $y = 5x^2$, $y = 2x^2$, $y = 0.3x^2$, $y = -x^2$, *and* $y = -3x^2$. Which graph represents the following equations?

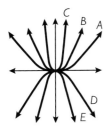

7. $y = 0.3x^2$

8. $y = 5x^2$

9. $y = -x^2$

10. $y = 2x^2$

11. $y = -3x^2$

Extending Concepts

Tell whether each statement is true or false. If it is true, explain why it makes sense. If it is false, change it so that it is true.

12. If the point $(3, 80)$ lies on the graph of an equation of the form $y = ax^2$, then the point $(-3, 80)$ also lies on the graph.

13. The graph of $y = 10x^2$ is wider than the graph of $y = 9x^2$.

14. If *a* is negative, the graph of $y = ax^2$ has no points above the *x*-axis.

15. The graph of $y = ax^2$ is symmetric about the *x*-axis.

16. You can get the graph of $y = -2x^2$ by flipping the graph of $y = 2x^2$ over the *x*-axis.

Writing

17. Answer the Dr. Math letter.

Dear Dr. Math,

I'm confused. I was trying to figure out the slope of the graph of $y = x^2$. I know the slope is the amount that the graph rises when you move one unit to the right. When I moved one unit to the right from $x = 1$ to $x = 2$, the rise was 3. So I figured the slope must be 3. But when I moved one unit to the right from $x = 2$ to $x = 3$, the rise was 5. What's going on? What is the slope? Why did I get two different answers for the rise?

Going 'Round The Bend

Fenced In

Applying Skills

For each table of values, (a) make a graph; (b) describe the graph and find its line of symmetry; (c) find the value of x where the graph reaches its maximum height.

1.

x	y
0	0
1	10
2	16
3	18
4	16
5	10
6	0

2.

x	y
0	0
1	12
2	20
3	24
4	24
5	20
6	12
7	0

Extending Concepts

3. Suppose you want to find two positive numbers whose sum is 11 and whose product is as large as possible.

a. List all the whole-number possibilities for the two numbers. For each one, find the product. Organize your results in a table like the one shown.

1st Number	2nd Number	Product
1	10	10
2	9	18
⋮	⋮	⋮

b. Make a graph. Show the value of the first number on the horizontal axis and the product on the vertical axis.

c. According to your graph, what value for the first number results in the largest product?

Making Connections

4. If a ball is thrown straight up at a speed of 96 feet per second, its height (in feet) after t seconds will be $h = 96t - 16t^2$.

a. Find the height of the ball after 0, 1, 2, 3, 4, 5, and 6 seconds. Show your results in a table. (For example, the height after 2 seconds is $96 \cdot 2 - 16 \cdot 2^2 = 192 - 16 \cdot 4 = 192 - 64 = 128$ feet).

b. Use your table to plot points and make a graph. Show time on the horizontal axis and height on the vertical axis.

c. According to your graph, when does the ball reach its maximum height?

d. Use your graph to estimate the times at which the height of the ball is 100 feet. Why do you think there are two such times?

e. How could you have predicted from the equation that the parabola would open downward?

Folds and Regions

Homework 10

Applying Skills

1. Make a table of values for the function $y = 6^x$. Include $x = 0, 1, 2, 3, 4, 5$. Plot the points and make a graph of the function.

Suppose that a is greater than 1. Tell whether each graph below could be the graph of the function $y = a^x$. Explain why or why not.

2.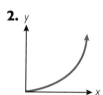

3.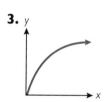

4.

5.

Tell whether each statement is true or false. If it is false, change it so that it is true.

6. The graph of $y = 9^x$ rises more steeply than the graph of $y = 8^x$.

7. The y-intercept of the graph of $y = 7^x$ is a higher point on the y-axis than the y-intercept of the graph of $y = 5^x$.

8. The y-intercept of the graph of $y = 9^x$ is $(0, 9)$.

9. The graph of $y = 8^x$ passes through the point $(1, 8)$.

Extending Concepts

10. a. Which is larger, x^2 or 2^x if x is 3? 8? 20? 50?

b. Which graph rises more steeply, $y = x^2$ or $y = 2^x$? Explain your thinking.

11. On April 1, Kate receives $10,000. At the end of the month, this amount is squared. On April 1, John receives $1. The next day and each of the following days until the end of the month, his money is doubled.

a. Predict who will have the most money at the end of the month and estimate how much more this person will have.

b. Using an exponent, write the amount of money each person will have at the end of the month.

c. Use your calculator to calculate how much money each person will have at the end of the month. Who will have more money? How much more?

Making Connections

12. Under favorable laboratory conditions, the number of cholera bacteria in a colony can double every half hour. If the colony starts with 1 bacterium, how many bacteria will there be at the end of 12 hours?

Rules of the Road for Exponents

Applying Skills

1. Make tables of values for the functions $y = 2^x$ and $y = 3^x$. Include the values $x = 0, 1, 2, ..., 12$.

Use your tables from item 1 to help you evaluate each expression. Then translate your result back into an expression using an exponent.

2. $2^3 \cdot 2^5$ **3.** $3^2 \cdot 3^7$ **4.** $2^6 \cdot 2^5$

5. $(2^3)^4$ **6.** $(3^2)^5$ **7.** $(2^4)^2$

8. $\dfrac{2^9}{2^4}$ **9.** $\dfrac{3^7}{3^6}$

Use the laws of exponents to simplify each expression. Write your answer using an exponent.

10. $4^3 \cdot 4^8$ **11.** $(x^5)^3$ **12.** $\dfrac{2^{14}}{2^2}$

13. $\dfrac{5^{16}}{5^9}$ **14.** $(8^4)^9$ **15.** $a^7 \cdot a^9$

16. $5^6 \cdot 5^{10}$ **17.** $(a^3)^{10}$ **18.** $\dfrac{6^{18}}{6^{15}}$

19. $x^8 \cdot x^{20}$ **20.** $(3^7)^2$ **21.** $\dfrac{x^{21}}{x^{15}}$

Evaluate each expression. Write your answer as a fraction.

22. 2^{-3} **23.** 3^{-2} **24.** 2^{-5}

25. 4^{-3} **26.** 5^{-4}

Extending Concepts

Tell whether the second expression is greater than, less than, or equal to the first expression.

27. $90^0, 1^{90}$

28. $3^0, 10^{-1}$

29. $20^3, 3^{20}$

30. $2^{-4}, 4^{-2}$

Making Connections

31. The half-life of a radioactive substance is the time it takes for half the amount originally present to decay. If the half-life of a particular substance is 1 year and if 1 gram is originally present, the amount remaining after x years will be 2^{-x} grams.

a. How much will remain after 5 years? after 8 years?

b. Will the amount remaining ever reach zero? How do you know?

The Very Large and the Very Small

Applying Skills

Write each number using scientific notation.

1. 8,200   **2.** 870,000

3. 2,500,000   **4.** 76,000

5. 0.03   **6.** 0.00064

7. 0.0014   **8.** 0.00000007

Write each number using standard notation.

9. 4.7×10^4 **10.** 1.9×10^7

11. 6×10^2 **12.** 8.5×10^{10}

13. 7×10^{-2} **14.** 3.3×10^{-5}

15. 3×10^{-8} **16.** 9.2×10^{-11}

Use your calculator to multiply each pair of numbers. Give your answer in scientific notation and in standard notation.

17. 23,000 and 11,400,000

18. 0.000041 and 0.000006

Extending Concepts

19. The distance from the planet Mercury to the sun is about 3.7×10^7 miles. The distance from Pluto to the sun is about 3.7×10^9 miles.

a. Estimate how much greater the distance from Pluto to the sun is than the distance from Mercury to the sun.

b. Calculate how much further it is to the sun from Pluto than from Mercury by writing the distances in standard notation and subtracting. Give your answer in scientific notation and standard notation.

Writing

20. Answer the Dr. Math letter.

> Dear Dr. Math,
>
> I've found a great method for converting a number into scientific notation. I just count the number of zeros and that's my exponent. So 24,000 is 24×10^3 because it has 3 zeros. 32,800,000 is 328×10^5 because it has 5 zeros. My method even works for the small numbers: 0.0005 has 3 zeros after the decimal point so it's 5×10^{-3}. How do you like my method? Should I tell my teacher about it?
>
> Lucy

Glencoe

This unit of MathScape: Seeing and Thinking Mathematically was developed by the Seeing and Thinking Mathematically project (STM), based at Education Development Center, Inc. (EDC), a non-profit educational research and development organization in Newton, MA. The STM project was supported, in part, by the National Science Foundation Grant No. 9054677. Opinions expressed are those of the authors and not necessarily those of the Foundation.

CREDITS: Photography: Chris Conroy • © Tom Tracy/Tony Stone Images: pp. 269TL, 278 • Kathleen Campbell/Tony Stone Images: pp. 269TC, 286.

Send all inquiries to:
Glencoe/McGraw-Hill
8787 Orion Place
Columbus, OH 43240-4027

ISBN: 0-07-866832-8

5 6 7 8 9 10 058 06